POWERBOATS

BY DENNY VON FINN

TORQUE™

BELLWETHER MEDIA • MINNEAPOLIS, MN

Are you ready to take it to the extreme? Torque books thrust you into the action-packed world of sports, vehicles, and adventure. These books may include dirt, smoke, fire, and dangerous stunts. WARNING: read at your own risk.

This edition first published in 2010 by Bellwether Media, Inc.

Library of Congress Cataloging-in-Publication Data

Von Finn, Denny
Powerboats / by Denny Von Finn.
p. cm. – (Torque. The world's fastest)
Includes bibliographical references and index.
Summary: "Amazing photography accompanies engaging information about powerboats. The combination of high-interest subject matter and light text is intended for students in grades 3 through 7" –Provided by publisher.
ISBN 978-1-60014-282-6 (hardcover : alk. paper)
1. Motorboat racing--Juvenile literature. 2. Motorboats--Juvenile literature. I. Title.
GV835.9.V6 2009
623.82'31--dc22

2009013761

Printed in the United States of America.

CONTENTS

What Are Powerboats? 4
Powerboat Technology 10
The Future of Powerboats 16
Glossary 22
To Learn More 23
Index 24

What Are Powerboats?

Powerboats are boats with huge engines. **Pleasure boats** are powerboats people use for fun. Other powerboats are designed for competitive racing. Both kinds of powerboats have engines that can push them over 150 miles (241 kilometers) per hour.

SCREAM MACHINE
FOUNTAIN
J.E. EDWARDS
GRIFFIN POWER
Panasonic
SEA TOW
MOREHEAD CITY, NC
P2-9
FOUNTAIN
SHOGREN
Panasonic
76
76
GREN
gofountain.com
PERFORMANCE
LUCAS
OIL PRODUCTS INC.
7
7
DoubleTree
Grand Key Resort

The world's fastest pleasure boats are more than 30 feet (9 meters) long. The fastest have one or two motors. The motors are much like high-performance race car engines. One of these engines alone can produce more than 500 **horsepower**!

VELOCITY
29

There are many types of racing powerboats. These boats can have **inboard** or **outboard** motors. Hydroplanes are the fastest racing powerboats. Boats in the **Unlimited** hydroplane class reach speeds over 200 miles (322 kilometers) per hour!

Fast Fact

Ken Warby drove his jet-powered hydroplane to a world speed record of nearly 318 (512 kilometers) miles per hour in 1978.

Powerboat Technology

Inboard and outboard motors have a **propeller**. The propeller is completely or partly underwater. The motor spins the propeller, which pushes the boat forward.

The most popular hydroplanes use **turbine** helicopter engines to spin their propellers. These heli-copter engines produce 2,650 horsepower. That's more than 12 times the power of a normal car engine!

FORMULA
FORMULABOATS.CO

Hydroplanes are designed to reduce **friction**. Friction is the resistance created when two objects rub together. Friction decreases a boat's speed. A hydroplane's **hull** is designed so that it skims the water surface. This reduces the friction between the hull and the water. It is called "planing."

AHERN Rentals
10
EMCOR
AHERN Rentals

The world's fastest powerboats use jet engines. A jet engine **compresses** air and mixes it with fuel. The fuel mixture is lit on fire. This causes the fuel mixture to expand. It rushes out the rear of the engine. The force of the rushing fuel mixture pushes the powerboat forward.

The Future of Powerboats

New materials and technology are changing powerboats. Powerboats used to be made out of wood. Boat builders began using lighter and stronger materials in the 1940s. One of these materials was **fiberglass**. Even stronger materials are used to build hydroplane hulls today. **Composites** are lightweight and strong. They also protect drivers well in a crash.

U-100
MIRAGE
MIRAGEBOATS.com
It's all about Water!

The way powerboats get their power is also changing. Boats powered by water jets are becoming popular. A water jet works on the same idea as a jet engine. A pump sucks up water. The water is quickly pushed out the back of the boat. This pushes the boat forward.

Fast Fact

Boats with water jets have no propellers. They can ride on as little as 2 inches (5 centimeters) of water!

60
Parker
ENZED

Fast Fact

The crew of Earthrace were attacked by pirates in their first attempt to travel around the globe in 2007!

Powerboat motors use a lot of gasoline. Boats powered by **biodiesel** have started winning races. Biodiesel is much cheaper and cleaner than gasoline. It does not harm wildlife if it is spilled in the water. The powerboat Earthrace raced around the globe in just 61 days in 2008. The 1,080-horsepower craft made the trip entirely on biodiesel!

GLOSSARY

biodiesel—a fuel made out of plants or the oils from plants

composites—hard manmade fabrics coated in plastic

compresses—squeezes

fiberglass—a strong fabric coated in a hard plastic-like substance

friction—a force caused by two objects rubbing together

horsepower—a unit for measuring the power of an engine

hull—the body of a boat

inboard—a boat motor that is located inside the boat's hull

outboard—a boat motor that is visible at the rear of the boat

pleasure boat—a boat used for recreational purposes, not racing

propeller—a bladed device that a motor turns to push a boat forward

turbine—an engine that uses turning blades to create power

Unlimited—the most powerful class of competition hydroplanes

TO LEARN MORE

AT THE LIBRARY

Dubowski, Mark. *Superfast Boats*. New York, N.Y.: Bearport Publishing, 2006.

Strobel Dieker, Wendy. *Hydroplanes*. Mankato, Minn.: Capstone Press, 2007.

Von Finn, Denny. *Hydroplanes*. Minneapolis, Minn.: Bellwether Media, 2008.

ON THE WEB

Learning more about powerboats is as easy as 1, 2, 3.

1. Go to www.factsurfer.com.
2. Enter "powerboats" into the search box.
3. Click the "Surf" button and you will see a list of related Web sites.

With factsurfer.com, finding more information is just a click away.

INDEX

biodiesel, 21
composites, 16
Earthrace, 20, 21
engines, 4, 6, 11
fiberglass, 16
friction, 13
fuel, 15
gasoline, 21
horsepower, 6, 11, 21
hull, 13, 16
hydroplanes, 8, 9, 11, 13
inboard motors, 8, 10
jet engines, 8, 15, 18
motors, 6, 8, 10
outboard motors, 8, 10
pirates, 20
planing, 13
pleasure boat, 4, 6
propeller, 10, 18
racing, 4, 21
Unlimited hydroplanes, 8
Warby, Ken, 9
water jet, 18

The images in this book are reproduced through the courtesy of: Paul Kemiel, front cover, pp. 8, 9, 11, 12-13, 14-15, 15 (lower), 16-17; Handout / Getty Images, pp. 4-5; Velocity Boats, pp. 6-7; Nik Taylor / Alamy, p. 10; Barry Bland / Alamy, pp. 18-19; Dual Aspect / Alamy, pp. 20-21.